目 次

1、空き缶圧縮ビジネス　空き缶圧縮器具の使用方法--------------------2

2、空き缶圧縮器具の使用方法と企画------------------------------3
　⑴　使用方法の特徴--3
　⑵　器具に付帯できて使用方法バラエティー--------------------4
　⑶　保管の仕方・特徴--4
　⑷　リサイクル工程での特徴----------------------------------5
　⑸　本器具構造の利用企画------------------------------------5
　⑹　空き缶圧縮器具の課題------------------------------------6

3、公報解説　空き缶圧縮器具・特許第４５０４４６０号----------7
　⑴　要約・課題--7
　⑵　解決手段--7
　⑶　特許請求の範囲--7
　⑷　発明の詳細な説明--8
　⑸　技術分野--8
　⑹　背景技術--8
　⑺　先行技術文献--8
　⑻　発明が解決しようとする課題------------------------------9
　⑼　課題を解決するための手段-------------------------------10
　⑽　発明の効果---11
　⑾　図面の簡単な説明---------------------------------------11
　⑿　発明を実施するための形態-------------------------------12
　⒀　産業上の利用可能性-------------------------------------22
　⒁　符号の説明---22
　⒂　図面の簡単な説明---------------------------------------22

4、英文--28

1、空き缶圧縮ビジネス　空き缶圧縮器具の使用方法

　　空き缶を圧縮するのに足で踏んで潰さない。空き缶を器具にセットして、ハンドルを回して、搾って圧縮する。

　　わずかな力で圧縮できるので、男女老若問わず、体の不自由な方でも、空き缶圧縮のビジネスができる。

　　小さな音で静かに空き缶を圧縮でき、回収した空き缶が、嵩張らないで蓄積できる。

　　空き缶回収ボックスに取り付ければ、嵩張らないから大量に蓄積できて、経済的効果が発生する。

　　「圧縮」とは爪が缶胴を押して星型に凹ます事で、「絞る」とはその空き缶の上下を異なる方向へ回して缶胴を縦方向に短くする事。両者を合わせて「圧縮」とも言う。

Recycling can business

How to use an empty can compressor implement

Although a can is compressed, it does not crush on foot.

A can is set to an instrument, and a handle is turned and compressed.

Since the body can compress by slight power even in the more inconvenient one, business is made.

A can can be compressed without making sound, and it can accumulate without being bulky in a box.

If it attaches to a can box, since it is not bulky, it can accumulate in large quantities, and an economical effect will occur.

2、空き缶圧縮器具の使用方法と企画

(1)【使用方法の特徴】

空き缶圧縮器具	従来の足踏み式空缶潰し機
① 空き缶をセットした器具を手で搾って圧縮する	・ 空き缶を挟んだ器具を足で踏んで潰す
② 体重を掛ける必用がないため自由な姿勢で圧縮作業ができる。そのため、乗車姿勢で圧縮可能、従って同乗者なら走行中も圧縮可能。	・ 乗車姿勢では、器具に体重を掛けられないため、使用できない。
③ 身体に起因する理由で立つ事のできない人でも、手だけで空缶を圧縮できる。	・ 立つ事のできない人は、空缶を潰せない。
④ 身体障害者に仕事を提供できる。※飲食会へ積極的な参加を促す。	・ 健常者の仕事である。
⑤ 爪が空缶の中心部へ移動して、梃子の応用で圧縮が軽くなる。但しサイズの違う空き缶に併用するには、回転止めの移動が必要。	・ アルミ缶に体重を掛けて潰す場合、器具を使用するほどの必要性はない。
⑥ 圧縮作業が安全で、結果が均一である。	・ 作業時に足を挫く事故があり、潰し結果も不揃いになり易い。
⑦ 飲食中に空になった缶を順次圧縮できるため、卓上が整理されて食材が際立ち、食後の片づけも軽減される。	・ 飲食の終了を待って潰す事になるため、飲食中は卓上に空缶が散乱する場合があり、片付けるときの負担も大きい。
⑧ 器具の爪で空き缶を圧縮する動作は、モーターの使用が可能。	・ 圧力を用いるため、大きさと重量を無視すれば、様々な動力の利用が可能。

（⑧の補足）特許図面のまま試作した場合、使用時に器具と空き缶の滑りを止めるように指を使用する、若しくは確実な滑り止めを取り付ける必要があるが、爪の動きを電動にするとこれらが解消する。

(2)【器具に付帯できて使用方法バラエティー】

空き缶圧縮器具	従来の足踏み式空缶潰し機
・ 時計　　・ キッチンタイマー ・ 計算機　・ 簡単な掲示板 ・ 公告　　・ 数量カウンター ・ カレンダー　・ 湿度計 ・ 室温計　・ 鏡 ・ カメラ　・ 翻訳機 ・ レシピ　・ ボイスレコーダー ・ メジャー　・ メモ ・ 聴覚障害者のための発光器 ・ 電動圧縮用モーター ・ 電子ゲーム（ランダム表示による運勢占い等） ・ 開栓器具（硬いペットボトルのキャップ開け） 　※この場合、器具の回転方向が指定される	・ 通常は掃除用具入れ等に収納しておくため、器具への付帯は意味を成さない。

※　特許図では説明の都合上ハンドルを円形に描画しているが、試作では手で回す事を考慮する。
　そのためハンドルの形状は特許図面より少し厚めで、角を丸めた四角形になる可能性が高い。
　また、キャラクターの輪郭を模った形状にする事でファンの人気を集める事も一案である。

(3)【保管の仕方・特徴】

空き缶圧縮器具	従来の足踏み式空缶潰し機
①　食器に順ずる物として位置付けできるので、食器棚に保管できる。そのため使用開始が容易。	・ 床に置いて足（履物）で踏むため、食器と同列には保管できない。使用後も手洗いが必用。
②　車のダッシュボードに保管でき、運転中に呑んだ空き缶をすぐ圧縮できる。	・ 車に常備するならトランクになり、使用は車外へ出て行う。
③　殆どが樹脂製で軽く、合体時は金属部分（爪）が内部に収納されるので、他の食器を傷つけない。	・ 全体が金属製で重く、掃除用具等と同列に収納する必要がある。

⑷【リサイクル工程での特徴】

	空き缶圧縮器具	従来の足踏み式空缶潰し機
①	缶胴の中心部がねじれ、その上下に空洞が出来る。 ※ 捻って圧縮した空き缶数百個を一塊に圧縮すると、空き缶の上下に残った空洞と中央のねじれが絡み合ってインゴット状になるため、輸送と作業に適しており、コストを下げる。圧縮されていない空き缶が混在していてもインゴット状になる。	・空き缶の上面と底面を畳むように缶胴の側面へ折込むためペチャンコになる。 ※ ペチャンコに潰れた空き缶は、平らな金属なので、いくら圧縮しても絡まず、バラバラであり、輸送や作業にコストがかかる。圧縮されていない空き缶に混入すると、インゴット化を妨げる。
②	前記【⑵器具に付帯できるもの】によって、主婦や子供の興味をそそり、ダイニングや車内に定着させる事で、空き缶圧縮率の向上を図り、輸送コストの低減に！	・空き缶潰し機の持ち出しと作業に手間が係り、缶を潰す快感も数本で薄れ、面白みも無く、大量の缶を一度に潰す事は苦痛であるため、実際に使用される事は希である。
③	自販機に器具を取り付けても歩行者が空けた缶を圧縮する事は期待できない。然しドライブ中の車から捨てられる空き缶が圧縮されていて、スペースを節約できる。	・缶が空いて直ぐ潰す事例は皆無であり、自販機等に併設されたゴミ箱は直ぐ満杯になる。
④	自販機に器具を取付ける場合、ハンドルの上下面は公告スペース。	・自販機に取り付ける事は可能。

⑸【本器具構造の利用企画】

	空き缶圧縮器具	従来の足踏み式空缶潰し機
①	本器具は、構造上物を掴んで一定方向へ増幅した力で回す事ができる。そのため、水道管や硬く絞まったハンドルなどを回す作業の応用。	
②	硬く絞まったハンドル状の物を緩め、又は絞める。	
③	爪の先を刃に代えてドラム缶やパイプ等をカットする。	
④	建屋の解体現場等で梁や柱を掴ませて引き抜く、或いは引き倒す。	
⑤	駐車中の二輪車の車輪を掴んでロックし、盗難を防止する。	
⑥	浮力を持たせた本器具で海底ケーブルを浮かせて移動し、所定の位置で開放する（沈める）。	

空き缶圧縮器具（以後、器具）を製造（試作）する上での課題

缶の素材は主にスチールとアルミニウムが使用されている。
- スチール缶の胴径は５３mmが多く、アルミニウム缶と比較して硬く破れ難い。
- アルミニウム缶の胴径は６６mmが多く、スチール缶と比較して軟らかく破れ易い。
- ブリキ缶は中間の素材として扱う事が出来る。

器具の使用者は何れの缶製品も購入するので、何れにも対応する事で価値が高まる。一つの器具でサイズと性質の違う金属缶を圧縮するための課題は次の通り。

【１】缶胴のサイズに合わせて爪の移動ストロークを切替え可能にする。
- 空き缶を捻るために必要な爪の圧縮量（深さ）は、スチール缶の方が内側にあって大きなてこ比を得ることができ、缶の硬さに合っているため都合が良い。
- このとき爪が中央に接近するが、対向する爪同士は進路が延長線上からずれていて爪幅を制限しないため、アルミニウム缶を破らない広い爪幅を持たせたまゝで良い。
- 但し凹みの無い空き缶は空転し易いので、爪先に適度な滑り止めを持たせたい。従って爪先の要求は「滑らず破らず」であり材質はゴム系や樹脂系が好ましく検討が必用。

※ 爪が缶胴を突き抜けると圧縮しないため捻れない。無理に捻ると爪の側面が缶胴を切り破り、怪我の恐れがある。無理な使用を防止するには、爪が貫通した場合にハンドルを空回りさせる必用がある。その分コストが嵩む。

【２】使用者のための雫受けが必要。
- 缶飲料を飲んだ後に残る雫が、使用時に漏れて衣服を汚す恐れがある。但し、圧縮が完了すると中央の捻れで密閉される。そのため一時的な雫の吸収、若しくは密閉で良い。コストと持続性からゴム板等で一時的に開口部を塞ぐことで十分。

【３】食品用の空き缶（主にブリキ缶）を捻る。
- 食品用の空き缶は蓋が取り去られており、爪の圧縮で開口部が変形するため圧縮できない。然し器具側に開口部の形状を維持する円形の壁を付ければ圧縮可能。壁は開口部を内側から支えれば良く、不連続でも機能するため飲料缶の圧縮には影響しない。

3、公報解説　空き缶圧縮器具・特許第４５０４４６０号

特許第４５０４４６０号
発明の名称；空き缶圧縮器具
特許権者／発明者；田中　勝之

⑴【要約】【課題】
　小さな力で空き缶を容易に圧縮することができるとともに、操作が簡単で耐久性に富む空き缶圧縮器具を提供する。
⑵【解決手段】回転ハンドル２を上側に配置して見たときに、回転ハンドル２の下側に接して回転ハンドル支持部３が設けられている。回転ハンドル支持部３に接して、回転ハンドル２とは反対側に、第一基板４、第二基板５、第三基板６、第四基板７が順次接するように設けられている。第二基板５には、２つの空き缶押圧部８ａが互いに対向するように取り付けられ、第三基板６には、２つの空き缶押圧部８ｂが互いに対向するように取り付けられている。回転ハンドル２を回転すると、可動部９が第二基板５と第三基板６の中心方向に向かってスライドする。その結果、空き缶圧縮爪１１は、中心方向に向かって移動し、空き缶圧縮爪１１の先端部１２が空き缶の外周に接触した後、空き缶を押圧して圧縮する。
【選択図】図１

⑶【特許請求の範囲】
【請求項１】
空き缶の上面側と底面側とに設置され、回転ハンドルと回転ハンドル支持部とが本体部に取り付けられて形成され、前記回転ハンドルは前記回転ハンドル支持部に対して回転可能である空き缶圧縮器具であって、前記回転ハンドルの中心を通る回転軸が、前記本体部の中心を貫通する部位において、前記回転軸の外周に接して歯車が設けられ、前記歯車の凹凸と噛み合う凹凸を有するラックギアが形成され、前記ラックギアに連動する可動部を有する空き缶押圧部は空き缶圧縮爪を備えており、前記回転ハンドルの回転が前記歯車に伝達されて、前記ラックギアを介して前記可動部の並進運動に変換され、前記空き缶押圧部の前記空き缶圧縮爪の先端部が、空き缶の中心方向に移動して、前記空き缶を押圧して圧縮することを特徴とする空き缶圧縮器具。
【請求項２】

前記回転ハンドルに接して前記回転ハンドル支持部が設けられ、前記回転ハンドル支持部は前記本体部に固定されており、前記回転ハンドルには回転止め壁部が設けられるとともに、前記回転ハンドル支持部には回転止めが設けられ、前記回転ハンドルが所定の角度だけ回転すると、前記回転止め壁部が前記回転止めに当接して、前記回転ハンドル支持部に対する前記回転ハンドルの回転が停止し、前記空き缶圧縮爪の動きを停止することを特徴とする請求項1記載の空き缶圧縮器具。

【請求項3】
一方の空き缶圧縮器具の回転ハンドルを上側にし、他方の空き缶圧縮器具の回転ハンドルを下側にして、2つの空き缶圧縮器具を合体したときに、上側に配置された空き缶圧縮器具の空き缶押圧部が収納される切込みが、下側に配置された空き缶圧縮器具に設けられるとともに、下側に配置された空き缶圧縮器具の空き缶押圧部が収納される切込みが、上側に配置された空き缶圧縮器具に設けられていることを特徴とする請求項1または2記載の空き缶圧縮器具。

⑷【発明の詳細な説明】
⑸【技術分野】
【0001】
本発明は、小さな力で簡単な操作により空き缶を容易に圧縮することができるため、空き缶を縮小化して効率的に廃棄することが可能な空き缶圧縮器具に関する。

⑹【背景技術】
【0002】
ビールやジュースなどの飲料が缶に入れられて多く販売されており、使用済みの空き缶の処理が大きな問題となっている。使用済みの空き缶をそのまま廃棄すると、廃棄物の容積が極めて大きくなるため、空き缶を圧縮して廃棄することが望ましい。しかし、使用済みの空き缶を手作業で圧縮して縮小化することは難しく、縮小化せずに空き缶を回収して、処理場で機械により圧縮作業を行うのが通常であった。また、使用者自身が空き缶を圧縮する際には、器具を用いずに手または足で作業を行うことになり、縮小化された空き缶の形状がまちまちで、形を揃えて重ねることができないため、空き缶の運搬が効率的に行われないという問題点があった。
空き缶を廃棄するために圧縮する器具の一例が、特許文献1、特許文献2に記載されている。

⑺【先行技術文献】
【特許文献】

【０００３】
【特許文献１】実開平５－２４１９３号公報
【特許文献２】特開平１０－１２８５９２号公報
【発明の開示】
⑻【発明が解決しようとする課題】
【０００４】
特許文献１に記載された空缶圧壊器具は、筒状の空き缶の上面および下面をそれぞれ支持する上支持体と下支持体とを有し、両支持体間を空き缶の背丈より幾分大きな距離にし、少なくとも２本のロープで両支持体を連結し、ロープには少なくとも１個の玉を配置した構成のものである。この空缶圧壊器具は、上支持体と下支持体との間に空き缶を挟み、両支持体を反対側に回転することにより、両支持体間を連結しているロープが空き缶に接するようになり、ロープにつけた玉が缶壁をロープのからまる斜め方向に押しつぶすものである。
【０００５】
しかし、空き缶を上下から挟みこむ上支持体と下支持体とは、ロープによって連結されているため、両支持体を反対側に回転したときに、ロープが両側から引っ張られることにより破断する恐れがある。また、ロープにつけられた玉が缶壁を押すことによって空き缶が圧縮されるため、空き缶を充分に圧縮するためには、ロープにつけられた玉が缶壁を強く押圧することが必要である。しかし、ロープに弛みが生じると、玉が缶壁を強く押圧することができない。
このように、特許文献１に記載された空缶圧壊器具では、玉が缶壁を強く押圧するようにするためには、ロープに弛みが生じないようにすることが必要であるが、そうすると、ロープは両側に引かれる力を強く受けるようになって、破断しやすくなるため、耐久性の点で問題が残る。
【０００６】
また、特許文献２に記載された空き缶つぶし技術は、空き缶の上面又は下面のいずれか一方を固定し、中心を軸とした回転力を与えつつ、中心軸と平行に、上面と下面に対して均一に加圧するものである。この技術においては、空き缶に対して中心軸のまわりに回転力を与えて捩じる動作が、空き缶つぶしのための基本動作となっているが、円筒形状が維持されたままの空き缶に対して回転力を与えて捩じるためには、かなりの力が必要であり、また、回転力を与える際の滑りが発生しやすいなどの問題点がある。
本発明は、このような事情を考慮してなされたもので、小さな力で空き缶を容易に

圧縮することができるとともに、操作が簡単で耐久性に富む空き缶圧縮器具を提供することを目的とする。
⑼【課題を解決するための手段】
【０００７】
以上の課題を解決するために、本発明の空き缶圧縮器具は、空き缶の上面側と底面側とに設置され、回転ハンドルと回転ハンドル支持部とが本体部に取り付けられて形成され、前記回転ハンドルは前記回転ハンドル支持部に対して回転可能である空き缶圧縮器具であって、前記回転ハンドルの中心を通る回転軸が、前記本体部の中心を貫通する部位において、前記回転軸の外周に接して歯車が設けられ、前記歯車の凹凸と噛み合う凹凸を有するラックギアが形成され、前記ラックギアに連動する可動部を有する空き缶押圧部は空き缶圧縮爪を備えており、前記回転ハンドルの回転が前記歯車に伝達されて、前記ラックギアを介して前記可動部の並進運動に変換され、前記空き缶押圧部の前記空き缶圧縮爪の先端部が、空き缶の中心方向に移動して、前記空き缶を押圧して圧縮することを特徴とする。

【０００８】
円筒形状が維持されたままの空き缶に対して、回転力を与えて捩じって空き缶を圧縮するにはかなりの力が必要であるが、本発明においては、空き缶押圧部の空き缶圧縮爪の先端部が、空き缶の中心方向に移動して、空き缶を押圧して空き缶の側面に凹みを生じさせたうえで、空き缶を捩じって圧縮することができるため、大きな力を必要としない。また、空き缶の側面に凹みを生じさせるための手段として、歯車とラックギアを用いているため、再現性良く簡単な操作で行うことができ、特許文献１に記載のもののように、耐久性の問題も生じない。さらに、空き缶の圧縮作業が終了した後、空き缶を取り除く際には、圧縮の際の逆の操作を行えばよく、空き缶の取り外しも容易である。

【０００９】
本発明においては、前記回転ハンドルに接して回転ハンドル支持部が設けられ、前記回転ハンドル支持部は前記本体部に固定されており、前記回転ハンドルには回転止め壁部が設けられるとともに、前記回転ハンドル支持部には回転止めが設けられ、前記回転ハンドルが所定の角度だけ回転すると、前記回転止め壁部が前記回転止めに当接して、前記回転ハンドル支持部に対する前記回転ハンドルの回転が停止し、前記空き缶圧縮爪の動きを停止するものであることが好ましい。

【００１０】
回転止め壁部が回転止めに当接するまでは、回転ハンドルは回転ハンドル支持部に

対して回転し、回転ハンドルの回転が回転軸に伝達されて歯車が回転し、ラックギアの動きに伴って空き缶押圧部が空き缶の中心方向に移動する。しかし、回転止め壁部が回転止めに当接すると、回転ハンドルは回転ハンドル支持部に対して回転できなくなり、歯車の回転が停止し、空き缶押圧部の動きが停止する。この状態で回転ハンドルを回すと、空き缶圧縮器具が一体として回転するようになるため、空き缶の上下に設置された２つの空き缶圧縮器具の回転ハンドルを、互いに異なる方向に回転すると、小さな力で効率良く空き缶を上下方向に圧縮することができる。このように、空き缶を圧縮するための操作を行うにあたって、回転止め壁部と回転止めを設けることによって、上述した異なる２つの動作を、回転ハンドルを回すという操作を連続して行うことで実現することができ、操作性の点で有利な構造を有しているということができる。
【００１１】
本発明においては、一方の空き缶圧縮器具の回転ハンドルを上側にし、他方の空き缶圧縮器具の回転ハンドルを下側にして、２つの空き缶圧縮器具を合体したときに、上側に配置された空き缶圧縮器具の空き缶押圧部が収納される切込みが、下側に配置された空き缶圧縮器具に設けられるとともに、下側に配置された空き缶圧縮器具の空き缶押圧部が収納される切込みが、上側に配置された空き缶圧縮器具に設けられていることが好ましい。
【００１２】
空き缶押圧部の空き缶圧縮爪は、径方向の中心に向かって突出する構造を有しているが、空き缶押圧部が収納される切込みを設けることによって、一方の空き缶圧縮器具の回転ハンドルを上側にし、他方の空き缶圧縮器具の回転ハンドルを下側にして、２つの空き缶圧縮器具を合体したときに、ほぼ円筒形状を維持して保管することができ、場所をとらず、一方を紛失することなく保管できる。
⑽【発明の効果】
【００１３】
本発明によると、小さな力で空き缶を容易に圧縮することができるとともに、操作が簡単で耐久性に富む空き缶圧縮器具を実現することができる。
⑾【図面の簡単な説明】
【００１４】
【図１】本発明の実施形態に係る空き缶圧縮器具の外観斜視図である。
【図２】本発明の実施形態に係る空き缶圧縮器具の外観斜視図である。
【図３】回転ハンドルの回転による、空き缶押圧部の動作のメカニズムを説明する

ための図である。

【図4】回転ハンドルを、回転ハンドル支持部と接する側から見た図である。

【図5】回転ハンドル支持部を、回転ハンドルと接する側から見た図である。

【図6】第一基板の平面図である。

【図7】第二基板の平面図である。

【図8】第三基板の平面図である。

【図9】第四基板の平面図である。

【図10】第四基板の平面図である。

【図11】空き缶圧縮器具の使用方法を説明するための図である。

【図12】空き缶圧縮器具の使用方法を説明するための図である。

【図13】空き缶圧縮器具の使用方法を説明するための図である。

【図14】空き缶圧縮器具の使用方法を説明するための図である。

【図15】不使用時の保管状態を示す図である。

⑿【発明を実施するための形態】

【0015】

以下に、本発明の空き缶圧縮器具を、その実施形態に基づいて説明する。

図1、図2は、本発明の実施形態に係る空き缶圧縮器具の外観斜視図である。

図1、図2に示すように、空き缶圧縮器具1は、回転ハンドル2を有しており、図1は、回転ハンドル2を上側にしたときの斜視図であり、図2は、回転ハンドル2を下側にしたときの斜視図である。空き缶圧縮器具1は、空き缶の上面側と底面側とにそれぞれ設置されて使用される。

【0016】

図1に示すように、回転ハンドル2を上側に配置して見たときに、回転ハンドル2の下側に接して回転ハンドル支持部3が設けられている。回転ハンドル2は回転ハンドル支持部3に対して回転可能である。回転ハンドル支持部3に接して、回転ハンドル2とは反対側に、第一基板4、第二基板5、第三基板6、第四基板7が順次接するようにして固定されている。回転ハンドル2、回転ハンドル支持部3、第一基板4、第二基板5、第三基板6、第四基板7は、軽くかつ変形しにくい材料を用いて形成することが好ましく、一例としてアクリル樹脂を用いることができる。ここでは、第一基板4、第二基板5、第三基板6、第四基板7が積層されて形成される構造体を本体部20と呼ぶ。そうすると、空き缶圧縮器具1は、回転ハンドル2と回転ハンドル支持部3が本体部20に取り付けられて形成されたものとなる。

【0017】

第二基板５には、２つの空き缶押圧部８ａが互いに対向するように取り付けられ、第三基板６には、２つの空き缶押圧部８ｂが互いに対向するように取り付けられている。

空き缶押圧部８ａは、第二基板５の径方向に対して可動である可動部９と、可動部９に対して垂直方向であって、第三基板６と第四基板７が積層された方向に延伸する腕に相当する延伸部１０と、延伸部１０に対して垂直方向に連続して設けられ、第四基板７から離れた位置において、第四基板７の中心に向かって突出する空き缶圧縮爪１１とを備えている。

同様に、空き缶押圧部８ｂは、第三基板６の径方向に対して可動である可動部９と、可動部９に対して垂直方向であって、第四基板７が積層された方向に延伸する腕に相当する延伸部１０と、延伸部１０に対して垂直方向に連続して設けられ、第四基板７から離れた位置において、第四基板７の中心に向かって突出する空き缶圧縮爪１１とを備えている。空き缶押圧部８ａ、８ｂは、金属によって形成されている。

【００１８】
空き缶圧縮爪１１の先端部１２、すなわち空き缶圧縮爪１１が空き缶に接触する部位は、平坦面であってもよく、あるいは上方から見たときにエッジが半円状をなす曲面としてもよい。空き缶圧縮爪１１の高さを揃えるために、すなわち、回転ハンドル２から空き缶圧縮爪１１までの距離を等しくするために、延伸部１０の長さは、空き缶押圧部８ａと空き缶押圧部８ｂとでは異なるようにしている。ここでは、空き缶押圧部８ａの延伸部１０の長さを、空き缶押圧部８ｂの延伸部１０の長さよりも長くしている。なお、延伸部１０の長さは、圧縮対象となる空き缶の形状や材質によって、適宜定められる。

【００１９】
第四基板７には、空き缶を配置したときに空き缶の位置ずれを防いで固定するための固定壁１３が設けられている。図１においては、固定壁１３は周方向に連続する壁によって形成され、図２においては、複数の壁を間隔置いて周方向に配置することにより固定壁１３を形成しているが、固定壁１３は空き缶の位置ずれを防止する機能を有していればよく、その形状は状況に応じて適宜選択され、さらにこれらの形状に限定されない。固定壁１３の内周側には、空き缶の滑り止めのための固定部５２が設けられている。この固定部５２については、図９、図１０に基づいて後に詳述する。

なお、以上の説明においては、複数の基板を積層することによって本体部２０を形成しているが、一体成型したものについて上記の構造を有するように加工してもよ

い。また、空き缶押圧部８ａと空き缶押圧部８ｂの数はこれに限定されず、状況に応じて適宜定められる。
【００２０】
図３に基づいて、回転ハンドルの回転による、空き缶押圧部の動作のメカニズムを説明する。
図３に示すように、回転ハンドル２の中心には回転軸１４が設けられており、回転軸１４は、回転ハンドル支持部３、第一基板４、第二基板５、第三基板６、第四基板７のそれぞれの中心を貫通して、座金１５を介して第四基板７側で固定されている。第二基板５と第三基板６の中心を貫通して、回転軸１４の外周に接して歯車１６が設けられており、歯車１６の長さは、第二基板５の厚みと第三基板６の厚みとの和に等しくなるようにしている。すなわち、回転ハンドル２の中心を通る回転軸１４には、本体部２の中心を貫通する部位において、回転軸１４の外周に接して歯車１６が設けられた構造となっている。歯車１６の径は、空き缶の大きさや材質に応じて適宜選択される。第四基板７の厚みは、空き缶１９を配置したときに、プルタブ１８と干渉しないことを考慮して設定される。
【００２１】
第二基板５に設けられた可動部９には、歯車１６の凹凸と噛み合う凹凸を有するラックギア１７が形成されている。同様に、第三基板６に設けられた可動部９にも、歯車１６の凹凸と噛み合う凹凸を有するラックギア１７が形成されている。
【００２２】
回転ハンドル２を回転すると、回転軸１４の回転に伴って歯車１６が回転し、歯車１６の回転運動がラックギア１７の並進運動に変換されて、可動部９が第二基板５と第三基板６の中心方向に向かってスライドする。その結果、空き缶押圧部８ａ、８ｂの空き缶圧縮爪１１は、可動部９と同様に、空き缶１９の中心方向に向かって移動し、空き缶圧縮爪１１の先端部１２が空き缶１９の外周に接触した後、空き缶１９を押圧して圧縮する。
【００２３】
また、回転ハンドル２をこれと反対方向に回転すると、回転軸１４の回転に伴って歯車１６が逆方向に回転し、歯車１６の回転運動がラックギア１７の並進運動に変換されて、可動部９が第二基板５と第三基板６の外周方向に向かってスライドする。その結果、空き缶押圧部８ａ、８ｂの空き缶圧縮爪１１は、可動部９と同様に、空き缶１９の外周方向に向かって移動し、圧縮された空き缶１９を取り出すことができる。

なお、図３では、歯車１６とラックギア１７との関係を説明する都合上、１つの平面内において、２つのラックギア１７が見えるように表示している。歯車１６とラックギア１７の具体的な構造については、図７、図８に基づいて後に詳述する。
【００２４】
図４は、回転ハンドル２を、回転ハンドル支持部３と接する側から見た図である。図４に示すように、回転ハンドル２の外周部には、回転ハンドル支持部３との噛み合わせ部２１が設けられている。回転ハンドル２の中心部には、回転軸１４の外周に接して回転軸受け２２が設けられ、回転軸受け２２と噛み合わせ部２１との間に、周方向に間隔を置いて複数の壁部２３ａが形成され、さらに、壁部２３ａよりも噛み合わせ部２１寄りに、壁部２３ａと同心円状に、周方向に間隔を置いて複数の壁部２３ｂが形成されている。壁部２３が形成されていることによって、回転ハンドル２の重量を軽量に維持しつつ、回転ハンドル２に上方から加えられる押圧力に対する強度を確保することができる。また、壁部２３ａ、２３ｂは、周方向に間隔を置いて設けられているため、回転ハンドル２を回転したときに、回転ハンドル支持部３との間の抵抗を小さくすることができる。
また、壁部２３ｂの外周側の一部に連続して、噛み合わせ部２１方向に延設された回転止め壁部２４が設けられている。
【００２５】
図５は、回転ハンドル支持部３を、回転ハンドル２と接する側から見た図である。図５に示すように、回転ハンドル支持部３の中心部には、回転軸１４を通すための穴２５が設けられている。回転ハンドル支持部３の外周部には、回転ハンドル２との噛み合わせ部２６が設けられており、噛み合わせ部２６の一部には、中心方向に向かって突出する回転止め２７が形成されている。回転止め２７が形成されていることにより、回転ハンドル２を所定の角度だけ回転したときに、図４に示す回転止め壁部２４が回転止め２７に当接して、回転ハンドル２の回転角度が適切な角度となったところで、回転ハンドル支持部３に対する回転ハンドル２の回転が停止して、空き缶圧縮爪１１の動きを停止することができる。
【００２６】
本発明の空き缶圧縮器具１では、回転ハンドル２を回転することにより、空き缶圧縮爪１１を移動させて空き缶１９の側面を押圧して凹みを生じさせた後、空き缶圧縮爪１１の動きを停止して、さらに空き缶１９を挟んだ状態で、上下の回転ハンドル２を互いに反対方向に回して空き缶１９を捩じることにより、空き缶１９を上下方向に圧縮するという動作が必要である。

【0027】
回転止め壁部24が回転止め27に当接するまでは、回転ハンドル2は回転ハンドル支持部3に対して回転し、回転ハンドル2の回転が回転軸14に伝達されて歯車16が回転し、ラックギア17の動きに伴って空き缶押圧部8a、8bが空き缶19の中心方向に移動する。しかし、回転止め壁部24が回転止め27に当接すると、回転ハンドル2は回転ハンドル支持部3に対して回転できなくなり、歯車16の回転が停止し、従って空き缶押圧部8a、8bの動きが停止する。この状態で回転ハンドル2を回すと、空き缶圧縮爪11が空き缶19の側面を変形させた状態で、1つの空き缶圧縮器具1は、回転ハンドル2と回転ハンドル支持部3と本体部20とが一体となって回転するようになる。この動作に基づいて、空き缶19の上下に配置された2つの空き缶圧縮器具1の回転ハンドル2を、互いに異なる方向に回転すると、小さな力でかつ単に回転ハンドル2を回転するだけの簡単な操作で、空き缶19を上下方向に圧縮することができる。

このように、空き缶19を圧縮するための操作を行うにあたって、回転止め壁部24と回転止め27を設けることによって、上述した異なる2つの動作を、回転ハンドル2を回すという操作を連続して行うことで実現することができ、操作性の点で有利な構造を有している。

【0028】
図6は、第一基板4の平面図である。

図6に示すように、第一基板4の中心部には、回転軸14を通すための穴31が設けられている。また、第一基板4の外周部から中心部に向かって、複数の切込み32が設けられている。この切込み32は、2つの空き缶圧縮器具1を不使用時に合体して保管しておくためのものである。

【0029】
不使用時には、一方の空き缶圧縮器具1の回転ハンドル2を上側にし、他方の空き缶圧縮器具1の回転ハンドル2を下側にして、2つの空き缶圧縮器具1を合体して保管しておくと、場所をとらず、一方を紛失することなく保存できる。空き缶圧縮爪11は、図1、図2に示すように、径方向の中心に向かって突出する構造を有しているため、切込み32は、2つの空き缶圧縮器具1を合体したときに、空き缶圧縮爪11を収納できるだけの深さを有するように形成されている。ここで、切込み32の深さとは、外周側から中心に向かって測ったときの切込み32の寸法のことをいう。2つの空き缶圧縮器具1を合体したときの状況は、図15を用いて後に詳述する。

【００３０】
図７は、第二基板５の平面図である。
図７に示すように、第二基板５の中心部には、回転軸１４の外周に接して歯車１６が設けられ、可動部本体４１には、歯車１６の凹凸と噛み合う凹凸を有するラックギア１７が形成されている。
回転ハンドル２を回転すると、回転軸１４の回転に伴って歯車１６が回転し、歯車１６の回転運動がラックギア１７の並進運動に変換される。ここでは、ラックギア１７によって２つの可動部本体４１が互いに反対方向にスライドし、可動部本体４１に連続して形成された可動部９が、第二基板５の中心方向に向かってスライドする並進運動を行う。その結果、空き缶押圧部８ａの空き缶圧縮爪１１は、空き缶１９の中心方向に向かって移動し、空き缶圧縮爪１１の先端部１２が空き缶１９の外周に接触した後、空き缶１９を押圧して圧縮する。
【００３１】
また、第二基板５の外周部には、４つの切込み４３が設けられている。この切込み４３は、２つの空き缶圧縮器具１を合体する際に、もう一方の空き缶圧縮器具１の空き缶押圧部８ａ、８ｂの延伸部１０を収納するためのものである。
【００３２】
図８は、第三基板６の平面図である。
図８に示すように、第三基板６の中心部には、回転軸１４の外周に接して歯車１６が設けられ、可動部本体４１には、歯車１６の凹凸と噛み合う凹凸を有するラックギア１７が形成されている。
回転ハンドル２を回転すると、回転軸１４の回転に伴って歯車１６が回転し、歯車１６の回転運動がラックギア１７の並進運動に変換される。ここでは、ラックギア１７によって２つの可動部本体４１が互いに反対方向にスライドし、可動部本体４１に連続して形成された可動部９が、第三基板６の中心方向に向かってスライドする並進運動を行う。その結果、空き缶押圧部８ｂの空き缶圧縮爪１１は、空き缶１９の中心方向に向かって移動し、空き缶圧縮爪１１の先端部１２が空き缶１９の外周に接触した後、空き缶１９を押圧して圧縮する。
【００３３】
このように、第三基板６は、第二基板５とほぼ同様の構造を有するものであり、第二基板５に対して９０度回転させて積層される。従って、回転ハンドル２を回転すると、第三基板６に取り付けられた可動部９は、第二基板５に取り付けられた可動部９に対して垂直な方向にスライドする。その結果、図１、図２に示した空き缶押

圧部８ａ、８ｂが、空き缶１９を四方から押圧するようになる。
第三基板６の外周部には、可動部本体４１に対して垂直な方向であって、第三基板６の中心部に向かって、２つの切込み４２が設けられている。この切込み４２は、図７に示す第二基板５に設けられた可動部９に連動して、空き缶押圧部８ａがスライドする際のスライドガイドとして機能する。
【００３４】
また、第三基板６の外周部には、可動部９と切込み４２との間に、４つの切込み４３が設けられている。この切込み４３は、２つの空き缶圧縮器具１を合体する際に、もう一方の空き缶圧縮器具１の空き缶押圧部８ａ、８ｂの延伸部１０を収納するためのものである。
【００３５】
図９、図１０は、第四基板７の平面図である。このうち、図９は、空き缶の上側に配置される空き缶圧縮器具１に用いられるものであり、図１０は、空き缶の下側に配置される空き缶圧縮器具１に用いられるものである。
図９、図１０に示すように、第四基板７の中心部には、回転軸１４を通すための穴５１が設けられている。また、第四基板７上には、空き缶を固定するための固定壁１３が形成されている。固定壁１３の内周寄りには、空き缶１９を配置して捩じったときに、空き缶１９が滑らずに安定的に固定されるための固定部５２が設けられている。固定部５２は、表面に微細な凹凸を設けることによって形成することができる。また、この部分を、滑り摩擦の大きい素材で形成してもよい。あるいは、空き缶１９を配置して捩じったときに、空き缶１９が滑らずに安定的に固定されるための手段であれば、他の構造のものであってもよい。空き缶は、上側で径が絞られているため、図９に示すものは、図１０に示すものよりも、固定壁１３を中心寄りに形成することにより、固定壁１３が空き缶の上側の側面を固定しやすいようにしている。
【００３６】
第四基板７の外周部から中心部に向かって、４つの切込み５３が設けられている。この切込み５３は、空き缶押圧部８ａ、８ｂがスライドする際のスライドガイドとして機能する。また、４つの切込み５３の間に、第四基板７の外周部から中心部に向かって、４つの切込み５４が設けられている。この切込み５４は、２つの空き缶圧縮器具１を不使用時に合体して保管しておく際に、もう一方の空き缶圧縮器具１の空き缶押圧部８ａ、８ｂを収納するためのものである。
【００３７】

図１１から図１４に基づいて、空き缶圧縮器具１の使用方法を説明する。
図１１は、一方の空き缶圧縮器具１の回転ハンドル２を下側にして配置し、この空き缶圧縮器具１上に空き缶１９の底面を固定し、空き缶１９の上端側に、もう一方の空き缶圧縮器具１を、回転ハンドル２を上側にして配置している様子を示している。
【００３８】
図１２は、２つの空き缶圧縮器具１の回転ハンドル２を、互いに異なる方向に回転している状況を示しており、回転ハンドル２の回転に伴って、空き缶押圧部８ａ、８ｂの空き缶圧縮爪１１が空き缶１９の側面を押圧する。その結果、空き缶１９は、押圧された部位が圧縮され、空き缶１９の側面に凹みが生じる。
【００３９】
図１３は、空き缶圧縮器具１によって空き缶１９の側面が圧縮される様子を、空き缶１９の上方から見た図であり、空き缶圧縮爪１１が空き缶１９の側面を押圧して、円筒形状であった空き缶１９に凹みが生じている。
２つの空き缶押圧部８ａの空き缶圧縮爪１１は、それぞれの延長線上から少し横方向にずれた位置で、空き缶１９に接触し、同様に、２つの空き缶押圧部８ｂの空き缶圧縮爪１１は、それぞれの延長線上から少し横方向にずれた位置で、空き缶１９に接触するため、空き缶１９の周方向に、規則的な凹みが形成される。空き缶１９の形状を、円筒形状からこのような凹みを有する形状とすることによって、この後の捩じりの工程では小さな力で済むようになる。
【００４０】
図１４は、空き缶１９の中心方向への空き缶押圧部８ａ、８ｂのスライドを停止した後に、さらに回転ハンドル２を持って空き缶１９を捩じった状態を示している。空き缶押圧部８ａ、８ｂのスライドの停止は、前述した、回転止め壁部２４が回転止め２７に当接することによって実現され、これを転換点として、２つの異なる動作を、回転ハンドル２の回転という一連の連続操作によって行うことができる。この捩じりの動作の際に、固定部５２が空き缶１９の上面と底面に接触するため、空き缶１９が空転することを防止でき、空き缶１９の上部は、上側に設置された空き缶圧縮器具１と一体となって回転し、空き缶１９の下部は、下側に設置された空き缶圧縮器具１と一体となって回転する。その結果、空き缶１９は上下方向に短時間で圧縮される。
また、空き缶１９の周方向に、規則的な凹みが形成された後に空き缶１９を捩じっているため、捩じられて上下方向に圧縮された空き缶１９は、綺麗な形に整えられ

ており、積み重ねて運搬する上で都合がよい。
【００４１】
空き缶１９を、円筒形状を保ったままで捩じって圧縮するためには、大きな力を必要とするが、空き缶圧縮器具１を用いると、空き缶１９の側面を押圧して凹みを生じさせた後、捩じる動作を行うため、大きな力を必要とせずに、効率良く空き缶１９を上下方向に圧縮することができる。空き缶１９の側面に凹みを生じさせる点については、特許文献１に記載のものと同じであるが、特許文献１に記載のものでは、ロープにつけた玉が缶壁をロープのからまる斜め方向に押しつぶすものであり、そのためには、ロープが弛みなく張っていることが必要であり、これによって耐久性の点で問題を生じる。これに対し、本発明の空き缶圧縮器具１は、空き缶押圧部８ａ、８ｂのスライドによって、空き缶１９の側面に凹みを生じさせており、安定的な動作が保証される。
【００４２】
また、特許文献１に記載のものでは、上支持体と下支持体との間に空き缶を挟み、両支持体を反対側に回転することによって、ロープにつけた玉が缶壁を押しつぶすものであるため、空き缶を圧縮するのに必要な力は結局、上支持体と下支持体とを捩じる力によって得られるのであり、比較的大きな力が必要である。これに対し、本発明の空き缶圧縮器具１では、回転ハンドル２の回転だけで、空き缶１９の側面に凹みを生じさせる動作と、空き缶１９を上下から捩じる動作を実現することができ、大きな力を必要としない。
【００４３】
また、回転ハンドル２の径は、空き缶１９の径より大きいため、回転ハンドル２を回転する力を小さくしても、大きな回転トルクを得ることができ、この点についても有利である。さらに、空き缶１９を圧縮するための操作をすべて回転ハンドル２の回転によって行うことができ、操作が簡単であるとともに、操作に必要な時間も極めて短時間ですむ。
また、圧縮された空き缶１９を取り外すためには、回転ハンドル２を逆方向に回転すればよく、この際の操作も簡単である。特許文献１に記載のものでは、空き缶が圧縮された後では、玉のついたロープが潰れた空き缶に絡まった状態となっており、空き缶を取出しにくい。
【００４４】
図１５は、不使用時の保管状態を示す。
図１５に示すように、一方の空き缶圧縮器具１ａの回転ハンドル２を上側にし、他

方の空き缶圧縮器具１ｂの回転ハンドル２を下側にして、２つの空き缶圧縮器具１を合体して保管しておくと、場所をとらず、一方を紛失することなく保存できる。空き缶押圧部８ａ、８ｂの空き缶圧縮爪１１は、径方向の中心に向かって突出する構造を有しているが、第一基板４に設けられた切込み３２は、図６に基づいて説明したように、２つの空き缶圧縮器具１を合体したときに、空き缶圧縮爪１１を収納できるだけの深さを有するように形成されているため、回転ハンドル２を適宜回転して、空き缶圧縮爪１１を切込み３２に収納することができる。そのため、２つの空き缶圧縮器具１を合体したときに、全体としてほぼ円筒形状をなすコンパクトな形状とすることができる。

【００４５】
第一基板４に設けられた切込み３２は、図６に基づいて説明したように、図７、図８に示す切込み４３よりも、中心方向に向かって深さが深くなっており、これによって中心方向に突出する空き缶圧縮爪１１の収納が可能となる。空き缶圧縮爪１１が切込み３２に収納されると、空き缶圧縮爪１１は、回転ハンドル支持部３と第二基板５とに挟みこまれるため、きっちりと固定される。

【００４６】
図１５では、上側に配置された空き缶圧縮器具１ａの空き缶押圧部８ａ１の空き缶圧縮爪１１が、下側に配置された空き缶圧縮器具１ｂの第一基板４に設けられた切込み３２に収納され、下側に配置された空き缶圧縮器具１ｂの空き缶押圧部８ａ２の空き缶圧縮爪１１が、上側に配置された空き缶圧縮器具１ａの第一基板４に設けられた切込み３２に収納されている。他の空き缶圧縮爪１１についても、同様の収納がなされている。これにより、空き缶１９を圧縮するにあたって、２つの空き缶圧縮器具１ａ、１ｂが必要であっても、不使用時には、これらを合体して一体物とすることができる。このような状態で保管できるため、家庭や飲食店、あるいは車内などにおいて、場所を取らずに保管でき、空き缶ができたときに手軽に使用することができる。このような保管状態にある空き缶圧縮器具１ａ、１ｂを使用したいときには、回転ハンドル２を回転すれば、切込み３２に収容されていた空き缶圧縮爪１１が外周方向にスライドして合体状態が解除される。

【００４７】
特許文献１に記載のものでは、上支持体と下支持体とがロープで連結されて一体物となっているが、不使用時には、必然的に、上支持体と下支持体との間で、玉が付随したロープが弛んだ状態で存在しており、定まった形状の定形物として保管することができない。そのため、保管場所から取出すときに、ロープが絡まったり、ロ

ープが他のものに引っかかったりして、使用上不便である。これに対し、本発明の空き缶圧縮器具1では、２つの空き缶圧縮器具1が対で使用されるにも拘わらず、２つの空き缶圧縮器具1を合体して、定形物として保管することができ、保管の際にも利便性が高い。

⒀【産業上の利用可能性】
【００４８】
本発明は、小さな力で空き缶を容易に圧縮することができるとともに、操作が簡単で耐久性に富む空き缶圧縮器具として利用することができる。特に、家庭や飲食店等において、ビールやジュースを飲んだ後に、使用済みの空き缶を廃棄するために、手軽に使用できる空き缶圧縮器具として利用することができる。

⒁【符号の説明】
【００４９】
１、１ａ、１ｂ　空き缶圧縮器具，２　回転ハンドル，３　回転ハンドル支持部，４　第一基板，５　第二基板，６　第三基板，７　第四基板，８ａ、８ｂ、８ａ１、８ａ２　空き缶押圧部，９　可動部，１０　延伸部，１１　空き缶圧縮爪，１２　先端部，１３　固定壁，１４　回転軸，１５　座金，１６　歯車，１７　ラックギア，１８　プルタブ，１９　空き缶，２０　本体部，２１　噛み合わせ部，２２　回転軸受け，２３、２３ａ、２３ｂ　壁部，２４　回転止め壁部，２５　穴，２６　噛み合わせ部，２７　回転止め，３１　穴，３２　切込み，４１　可動部本体，４２　切込み，４３　切込み，５１　穴，５２　固定部，５３　切込み，５４　切込み

⒂【図面の簡単な説明】
【００１４】
【図１】本発明の実施形態に係る空き缶圧縮器具の外観斜視図である。
【図２】本発明の実施形態に係る空き缶圧縮器具の外観斜視図である。
【図３】回転ハンドルの回転による、空き缶押圧部の動作のメカニズムを説明するための図である。
【図４】回転ハンドルを、回転ハンドル支持部と接する側から見た図である。
【図５】回転ハンドル支持部を、回転ハンドルと接する側から見た図である。
【図６】第一基板の平面図である。
【図７】第二基板の平面図である。
【図８】第三基板の平面図である。

【図９】第四基板の平面図である。
【図１０】第四基板の平面図である。
【図１１】空き缶圧縮器具の使用方法を説明するための図である。
【図１２】空き缶圧縮器具の使用方法を説明するための図である。
【図１３】空き缶圧縮器具の使用方法を説明するための図である。
【図１４】空き缶圧縮器具の使用方法を説明するための図である。
【図１５】不使用時の保管状態を示す図である。

【図１】

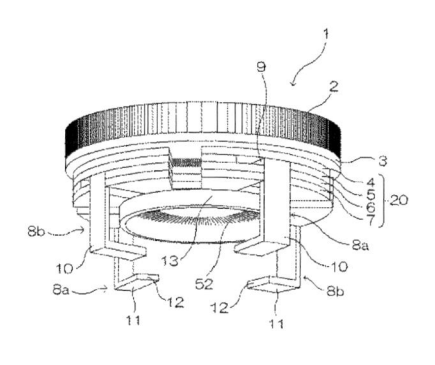

【図２】

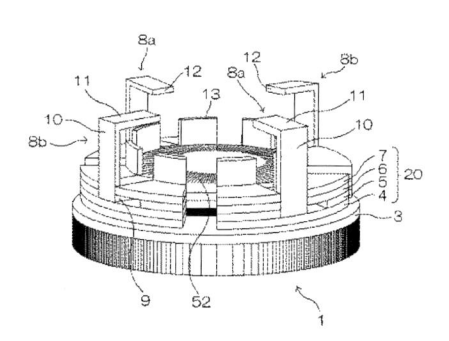

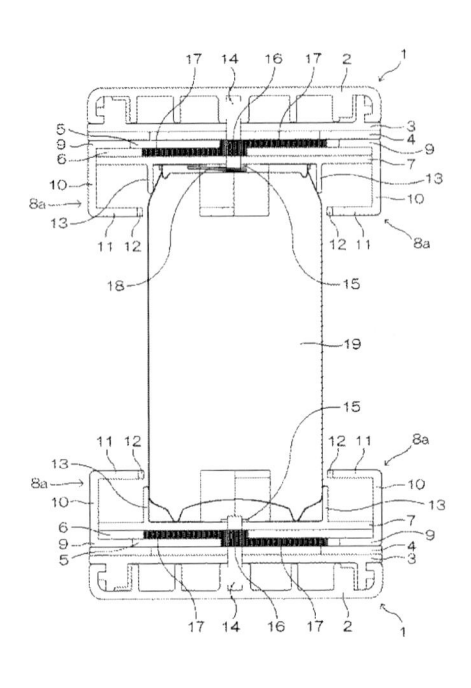

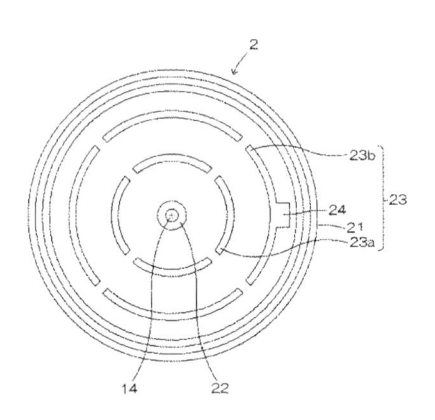

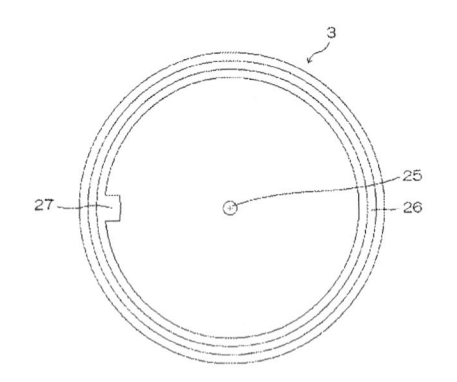

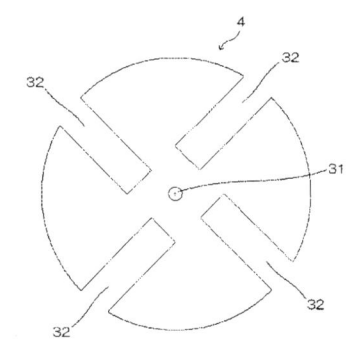

【図3】

【図4】

【図5】

【図6】

【図7】

【図8】

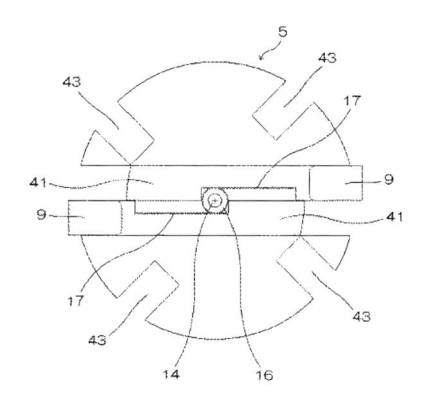

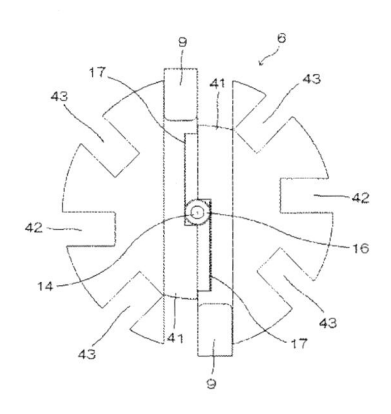

【図9】

【図10】

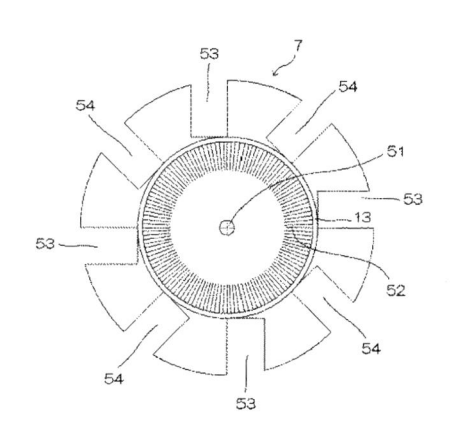

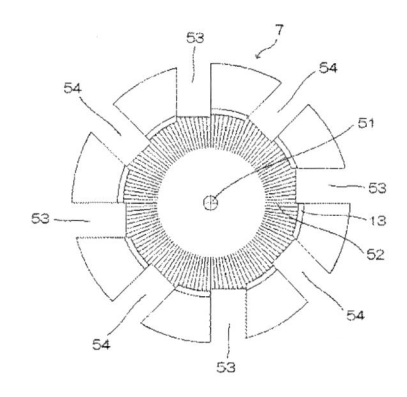

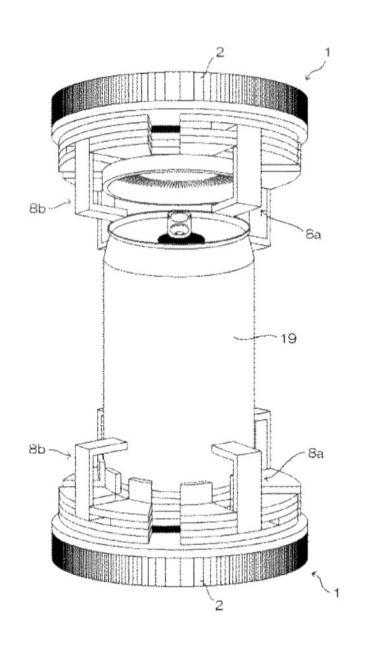

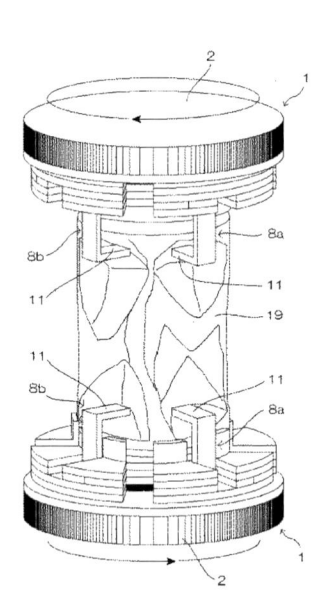

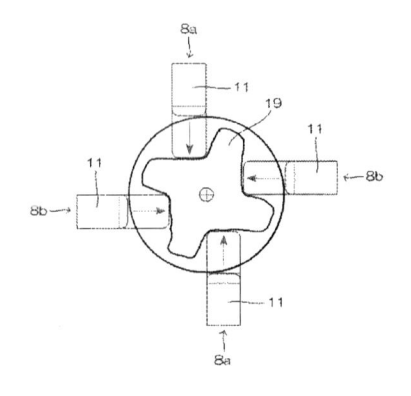

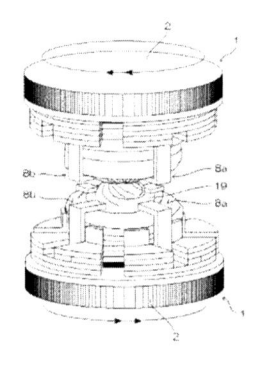

【図１１】

【図１２】

【図１３】

【図１４】

【図15】

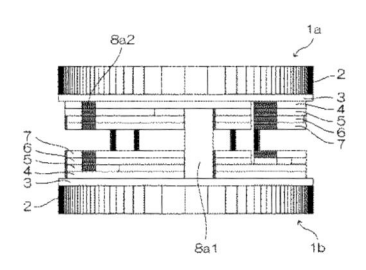

空き缶が十分圧縮されると、凸凹の缶胴は中心へ細く巻き込まれ、爪の停止位置より細くなって爪先から外れる。そのため、空き缶を絞り千切ることはない。

4、DETAILED DESCRIPTION

[Detailed Description of the Invention]

[Field of the Invention]

[0001]

Since the present invention can compress an empty can easily by easy operation by small power, it relates to the empty can compressor implement which reduction of the empty can is carried out and can discard it efficiently.

[Background of the Invention]

[0002]

It is put into beverages, such as beer and juice, by the can, they are sold mostly, and processing of a used empty can poses a big problem. Since the volume of waste will become very largely if a used empty can is discarded as it is, it is desirable to compress and discard an empty can. However, it was difficult to compress a used empty can manually and to carry out reduction, and empty cans is collected without carrying out reduction, and, usually the machine performed compression in the disposal plant. When the user itself compressed an empty can, it will work on a hand or foot, without using an instrument, the form of the empty can by which reduction was carried out was various, and since were able to arrange a form and it was not able to be piled up, there was a problem that conveyance of an empty can was not performed efficiently.

An example of the instrument compressed in order to discard an empty can is described in the Patent document 1 and the Patent document 2.

　[Citation list]

[Patent literature]

[0003]

[Patent document 1] JP,H5-24193,U

[Patent document 2] JP,H10-128592,A

[Description of the Invention]

[Problem to be solved by the invention]

[0004]

The empty can collapse instrument described in the Patent document 1 has the upper base material and lower base material which support the upper surface and the lower surface of a cylindrical empty can, respectively, It is a thing of composition of having made between both base materials into a bigger distance for how many minutes than the height of an empty can, having connected both base materials with at least two ropes, and having arranged at least one ball on the rope. By this empty can collapse instrument's sandwiching an empty can between an upper base material and a lower base material, and rotating both base materials to an opposite side, the rope which has connected between both base materials comes to touch an empty can, and the ball attached to the rope crushes a can wall to the oblique direction where a rope twines.

[0005]

However, since it is connected with the rope, the upper base material and lower base material which are full on both sides of an empty can from the upper and lower sides have a possibility of fracturing by pulling a rope from both sides, when both base materials are rotated to an opposite side. Since an empty can is compressed when the ball attached to the rope pushes a can wall, in order to compress an empty can sufficiently, it is required for the ball attached to the rope to press a can wall strongly. However, if slack arises on a rope, a ball cannot press a can wall strongly.

Thus, in the empty can collapse instrument described in the Patent document 1, in order for a ball to press a can wall strongly, it is required for a rope to keep slack from arising, but. Since a rope will come to receive strongly the power pulled at both sides and will become easy to fracture it if it does so, a problem remains on a point [durability].

[0006]

It pressurizes a medial axis and parallel uniformly to the upper surface and a lower surface, the empty can crushing technology described in the Patent document 2 fixing either one of the upper surface of an empty can, or a lower surface, and giving the rotational force centering on a center. Although the operation which gives and twists rotational force around a medial axis to an empty can in this technology is the basic motion for empty can crushing, In order for cylindrical shape to give and twist rotational force to the empty can maintained, remarkable power is required and there is a problem of being easy to generate the slide at the time of giving rotational force.

The the object of this invention is as follows.
It was made in consideration of such a situation, and an empty can can be easily compressed by small power.
Provide an empty can compressor implement which operation is easy and is rich in durability.

[Means for solving problem]
[0007]
In order to solve the above problem, the empty can compressor implement of the present invention, It is installed in the upper surface [of an empty can], and bottom surface side, and a rotating handle and a rotating handle supporting part are attached and formed in a body part, In the part to which the axis of rotation which the aforementioned rotating handle is a pivotable empty can compressor implement to the aforementioned rotating handle supporting part, and passes along the center of the aforementioned rotating handle penetrates the center of the aforementioned body part, The gear is provided in contact with the periphery of the aforementioned axis of rotation, and the rack gear which has unevenness of the aforementioned gear and the gearing unevenness is formed, The empty can pressing part which has a flexible region interlocked with the aforementioned rack gear is provided with the empty can compression nail, Rotation of the aforementioned rotating handle is transmited to the aforementioned gear, it converts to the translation motion of the aforementioned flexible region via the aforementioned rack gear, the point of the aforementioned empty can compression nail of the aforementioned empty can pressing part moves to the central direction of an empty can, and the aforementioned empty can is pressed and compressed.
[0008]
Although rotational force is given and twisted to the empty can by which cylindrical shape is maintained and remarkable power is required to compress an empty can, In the present invention, since an empty can can be twisted and compressed after the point of the empty can compression nail of an empty can pressing part moved to the central direction of an empty can, pressed an empty can and making the side surface of an empty can produce a dent, large force is

not needed. Since the gear and a rack gear are used as a means for making the side surface of an empty can produce a dent, it can carry out by the easy operation with sufficient reproducibility, and a durable problem is not produced like the thing of the description to a Patent document 1, either. When removing an empty can after compression of an empty can is completed, what is necessary is just to perform reverse operation in the case of compression, and removal of an empty can is also easy.

[0009]

While a rotating handle supporting part is provided in contact with the aforementioned rotating handle, the aforementioned rotating handle supporting part is fixed to the aforementioned body part in the present invention and a niting wall is provided by the aforementioned rotating handle, If niting is provided by the aforementioned rotating handle supporting part and the aforementioned rotating handle rotates only a predetermined angle, It is preferable that the aforementioned niting wall is what the aforementioned niting is abutted, and rotation of the aforementioned rotating handle to the aforementioned rotating handle supporting part stops, and stops a motion of the aforementioned empty can compression nail.

[0010]

It rotates to a rotating handle supporting part, and rotation of a rotating handle is transmited to the axis of rotation, the gear rotates, and an empty can pressing part moves a rotating handle to the central direction of an empty can in connection with a motion of a rack gear until a niting wall abuts niting. However, if a niting wall abuts niting, it will become impossible to rotate to a rotating handle supporting part, rotation of the gear will stop, and a motion of an empty can pressing part will stop a rotating handle. If the rotating handle of two empty can compressor implements installed in the upper and lower sides of an empty can is rotated in the mutually different direction in order for an empty can compressor implement to rotate as one, if a rotating handle is turned in this state, an empty can is efficiently compressible into a vertical direction by small power.

Thus, by providing a niting wall and niting in performing operation for compressing an empty can, It can realize by performing continuously operation of turning a rotating handle for two different operations mentioned above, and it can be said that it has an advantageous structure in respect of operativity.

[0011]

When the rotating handle of one empty can compressor implement is turned up, the rotating handle of the empty can compressor implement of another side is turned down in the present invention and two empty can compressor implements are united, While the notch by which the empty can pressing part of the empty can compressor implement arranged at the upper part is stored is provided by the empty can compressor implement arranged at the bottom, it is preferable that the notch by which the empty can pressing part of the empty can compressor implement arranged at the bottom is stored is provided by the empty can compressor implement arranged at the upper part.

[0012]

Although it has the structure projected toward the center of a radial direction, the empty can compression nail of an empty can pressing part, When the rotating handle of one empty can compressor implement is turned up, the rotating handle of the empty can compressor implement of another side is turned down and two empty can compressor implements are united by providing the notch by which an empty can pressing part is stored, Cylindrical shape can be maintained and kept substantially, a place is not taken, but it can be kept, without losing one side.

[Effect of the Invention]

[0013]

According to the present invention, while an empty can is easily compressible by small power, operation is easy and the empty can compressor implement which is rich in durability can be realized.

[Brief Description of the Drawings]

[0014]

[Drawing 1]It is an appearance perspective view of the empty can compressor implement concerning the embodiment of the present invention.

[Drawing 2]It is an appearance perspective view of the empty can compressor implement concerning the embodiment of the present invention.

[Drawing 3]It is a figure for describing the mechanism of operation of an empty can pressing part by rotation of a rotating handle.

[Drawing 4]It is the figure which looked at the rotating handle from the side which

touches a rotating handle supporting part.
[Drawing 5]It is the figure which looked at the rotating handle supporting part from the side which touches a rotating handle.
[Drawing 6]It is a plan view of the first substrate.
[Drawing 7]It is a plan view of the second substrate.
[Drawing 8]It is a plan view of the third substrate.
[Drawing 9]It is a plan view of the fourth substrate.
[Drawing 10]It is a plan view of the fourth substrate.
[Drawing 11]It is a figure for describing the directions for an empty can compressor implement.
[Drawing 12]It is a figure for describing the directions for an empty can compressor implement.
[Drawing 13]It is a figure for describing the directions for an empty can compressor implement.
[Drawing 14]It is a figure for describing the directions for an empty can compressor implement.
[Drawing 15]It is a figure showing the storage state at the time of non-use.
[Description of Embodiments]
[0015]
Below, the empty can compressor implement of the present invention is described based on the embodiment.
Fig.1 and Fig.2 are the appearance perspective views of the empty can compressor implement concerning the embodiment of the present invention.
As shown in Fig.1 and Fig.2, the empty can compressor implement 1 has the rotating handle 2, Fig.1 is a perspective view when the rotating handle 2 is turned up, and Fig.2 is a perspective view when the rotating handle 2 is turned down.The empty can compressor implement 1 is used for the upper surface [of an empty can], and bottom surface side, installing it, respectively.
[0016]
As shown in Fig.1, when the rotating handle 2 is arranged and seen to the up side, the rotating handle supporting part 3 is provided in contact with the rotating handle 2 bottom. The rotating handle 2 is pivotable to the rotating handle supporting part 3. The rotating handle supporting part 3 is touched, and in the

rotating handle 2, as the first substrate 4, the second substrate 5, the third substrate 6, and the fourth substrate 7 touch an opposite side sequentially, they are fixed to it. As for the rotating handle 2, the rotating handle supporting part 3, the first substrate 4, the second substrate 5, the third substrate 6, and the fourth substrate 7, forming using the material which cannot deform easily lightly is preferable, and an acrylic resin can be used for them as an example. Here, the first substrate 4, the second substrate 5, the third substrate 6, and the fourth substrate 7 call the structure laminated and formed the body part 20. When done so, the rotating handle 2 and the rotating handle supporting part 3 were attached to the body part 20, and the empty can compressor implement 1 was formed.
[0017]
It is attached so that the two empty can pressing parts 8a may oppose to the second substrate 5 mutually, and it is attached to the third substrate 6 so that the two empty can pressing parts 8b may oppose mutually.
The empty can pressing part 8a is provided with the following.
It is the movable flexible region 9 to a radial direction of the second substrate 5.
The extending part 10 equivalent to an arm extended in the perpendicular direction by which the third substrate 6 and the fourth substrate 7 were laminated to the flexible region 9.
The empty can compression nail 11 projected toward the center of the fourth substrate 7 in a position which was continuously provided perpendicularly to the extending part 10, and is separated from the fourth substrate 7.
Similarly, the empty can pressing part 8b is provided with the following.
It is the movable flexible region 9 to a radial direction of the third substrate 6.
The extending part 10 equivalent to an arm extended in the perpendicular direction by which the fourth substrate 7 was laminated to the flexible region 9.
The empty can compression nail 11 projected toward the center of the fourth substrate 7 in a position which was continuously provided perpendicularly to the extending part 10, and is separated from the fourth substrate 7.
The empty can pressing parts 8a and 8b are formed with metal.
[0018]
It may be a flat surface, or the part where the point 12 11 of the empty can compression nail 11, i.e., an empty can compression nail, contacts an empty can

has good edge also as a curved surface which makes semicircular state, when it sees from the upper part. In order to arrange the height of the empty can compression nail 11 (i.e., in order to make equal distance from the rotating handle 2 to the empty can compression nail 11), he is trying for the length of the extending part 10 to differ by the empty can pressing part 8a and the empty can pressing part 8b. Here, Sayori Cho of the extending part 10 of the empty can pressing part 8b also lengthens the length of the extending part 10 of the empty can pressing part 8a. The length of the extending part 10 is suitably defined by the form and construction material of an empty can used as a compression object.

[0019]

When an empty can has been arranged, the fixed wall 13 for preventing and fixing the positional displacement of an empty can is provided by the fourth substrate 7. Although the fixed wall 13 is formed with a circumferentially continuous wall in Fig.1 and the fixed wall 13 is formed in Fig.2 by arranging two or more walls to a between separate placement **** hoop direction,What is necessary is just to have the function to prevent the positional displacement of an empty can, the form is suitably chosen according to a situation, and the fixed wall 13 is not further limited to such form. The holding part 52 for the skid of an empty can is provided at the inner circumference side of the fixed wall 13. This holding part 52 is behind explained in full detail based on Fig.9 and Fig.10.

In the above description, although the body part 20 is formed by laminating two or more substrates, it may be processed so that it may have the above-mentioned structure about what was really molded. The number of the empty can pressing part 8a and the empty can pressing parts 8b is not limited to this, but is suitably defined according to a situation.

[0020]

Based on Fig.3, the mechanism of operation of the empty can pressing part by rotation of a rotating handle is described.

As shown in Fig.3, the axis of rotation 14 is provided by the center of the rotating handle 2, and the axis of rotation 14 penetrates each center of the rotating handle supporting part 3, the first substrate 4, the second substrate 5, the third substrate 6, and the fourth substrate 7, and is fixed by the fourth substrate 7 side via the washer 15. The center of the second substrate 5 and the third substrate 6 is penetrated, the gear 16 is provided in contact with the periphery of the axis of rotation 14, and he is trying for the length of the gear 16 to become equal to the sum of the thickness of the second substrate 5, and the thickness of the third substrate 6. That is, in the part which penetrates the center of the body part 2, it is the axis of rotation 14 which passes along the center of the rotating handle 2 with the structure where the gear 16 was provided in contact with the periphery of the axis of rotation 14. The path of the gear 16 is suitably chosen according to the size and construction material of an empty can. The thickness of the fourth substrate 7 is set up in consideration of not interfering with the pull tab 18, when the empty can 19 has been arranged.

[0021]

The rack gear 17 which has unevenness of the gear 16 and the gearing unevenness is formed in the flexible region 9 provided by the second substrate 5. Similarly, the rack gear 17 which has unevenness of the gear 16 and the gearing unevenness is formed also in the flexible region 9 provided by the third substrate 6.

[0022]

If the rotating handle 2 is rotated, the gear 16 will rotate with rotation of the axis of rotation 14, rotational movement of the gear 16 will be converted to the translation motion of the rack gear 17, and the flexible region 9 will slide toward the central direction of the second substrate 5 and the third substrate 6. As a result, the empty can compression nail 11 of the empty can pressing parts 8a and 8b presses and compresses the empty can 19, after it moves toward the central direction of the empty can 19 and the point 12 of the empty can compression nail 11 contacts the periphery of the empty can 19 like the flexible region 9.

[0023]

If the rotating handle 2 is rotated to this and a counter direction, the gear 16 will rotate to an opposite direction with rotation of the axis of rotation 14, rotational movement of the gear 16 will be converted to the translation motion of the rack gear 17, and the flexible region 9 will slide toward the outer peripheral direction of the second substrate 5 and the third substrate 6. As a result, like the flexible region 9, the empty can compression nail 11 of the empty can pressing parts 8a and 8b can be moved toward the outer peripheral direction of the empty can 19, and can take out the compressed empty can 19.

In Fig.3, it is displaying so that the two rack gears 17 can be seen on the convenience which describes the relation between the gear 16 and the rack gear 17, and in one plane. A specific structure of the gear 16 and the rack gear 17 is behind explained in full detail based on Fig.7 and Fig.8.

[0024]

Fig.4 is the figure which looked at the rotating handle 2 from the side which touches the rotating handle supporting part 3.

As shown in Fig.4, the engaging part 21 with the rotating handle supporting part 3 is provided by the outer peripheral part of the rotating handle 2. The axis-of-rotation receptacle 22 is provided in contact with the periphery of the axis of rotation 14 by the central part of the rotating handle 2, and between the axis-of-rotation receptacle 22 and the engaging part 21, An interval is kept circumferentially, two or more walls 23a are formed, further, rather than the wall 23a, an interval is circumferentially kept in the wall 23a and concentric circle shape, and two or more walls 23b are formed in engaging part 21 slippage. The strength to the thrust applied to the rotating handle 2 from the upper part is securable, maintaining the weight of the rotating handle 2 lightweight by forming the wall 23. Since an interval is kept circumferentially and it provides, the walls 23a and 23b can make small resistance of Hazama with the rotating handle supporting part 3, when the rotating handle 2 is rotated.

The niting wall 24 continuously installed in the part by the side of the periphery of the wall 23b in the engaging part 21 direction is provided.

[0025]

Fig.5 is the figure which looked at the rotating handle supporting part 3 from the side which touches the rotating handle 2.

As shown in Fig.5, the hole 25 for letting the axis of rotation 14 pass is provided by the central part of the rotating handle supporting part 3. The engaging part 26 with the rotating handle 2 is provided by the outer peripheral part of the rotating handle supporting part 3, and the niting 27 projected toward a central direction is formed in a part of engaging part 26 at it. When the rotating handle 2 is rotated only for a predetermined angle by forming the niting 27, the niting wall 24 shown in Fig.4 in the place where the niting 27 was abutted at and angle of rotation of the rotating handle 2 turned into a suitable angle, Rotation of the rotating handle 2 to the rotating handle supporting part 3 can stop, and a motion of the empty can compression nail 11 can be stopped.

[0026]

In the empty can compressor implement 1 of the present invention, by rotating the rotating handle 2, After having moved the empty can compression nail 11, pressing the side surface of the empty can 19 and producing a dent, where it stopped the motion of the empty can compression nail 11 and the empty can 19 is inserted further, By turning the up-and-down rotating handle 2 to a counter direction mutually, and twisting the empty can 19, operation of compressing the empty can 19 into a vertical direction is required.

[0027]

The niting wall 24 rotates the rotating handle 2 to the rotating handle supporting part 3 until it abuts the niting 27, rotation of the rotating handle 2 is transmited to the axis of rotation 14, the gear 16 rotates, and the empty can pressing parts 8a and 8b move to the central direction of the empty can 19 in connection with a motion of the rack gear 17. However, if the niting wall 24 abuts the niting 27, it will become impossible to rotate to the rotating handle supporting part 3, and rotation of the gear 16 will stop, therefore a motion of the empty can pressing parts 8a and 8b will stop the rotating handle 2. If the rotating handle 2 is turned in this state, in the state where the empty can compression nail 11 was made to deform the side surface of the empty can 19, the rotating handle 2, the rotating handle supporting part 3, and the body part 20 will be united, and the one empty can compressor implement 1 will come to rotate. If the rotating handle 2 of the two empty can compressor implements 1 arranged at the upper and lower sides of the empty can 19 is rotated in the mutually different direction based on this operation, it is small power and the empty can 19 can be compressed into a vertical direction by easy operation of only rotating the rotating handle 2.

Thus, in performing operation for compressing the empty can 19, by providing the niting wall 24 and the niting 27, it can realize by performing continuously operation of turning the rotating handle 2 for two different operations mentioned above, and has an advantageous structure in respect of operativity.

[0028]

Fig.6 is a plan view of the first substrate 4.
As shown in Fig.6, the hole 31 for letting the axis of rotation 14 pass is provided by the central part of the first substrate 4. Two or more notches 32 are provided toward the central part from the outer peripheral part of the first substrate 4. This notch 32 is for uniting at the time of non-use and keeping the two empty can compressor implements 1.

[0029]

At the time of non-use, the rotating handle 2 of one empty can compressor implement 1 is turned up, and the rotating handle 2 of the empty can compressor implement 1 of another side is turned down, and if it unites and the two empty can compressor implements 1 are kept, a place is not taken, but it can save, without losing one side. Since it has the structure projected toward the center of a radial direction as the empty can compression nail 11 is shown in Fig.1 and Fig.2, the notch 32 is formed so that it may have only the depth which can store the empty can compression nail 11, when the two empty can compressor implements 1 are united.Here, the depth of the notch 32 means the dimension of the notch 32 when it measures toward a center from the periphery side. When it unites, a situation explains the two empty can compressor implements 1 in full detail behind using Fig.15.

[0030]

Fig.7 is a plan view of the second substrate 5.
As shown in Fig.7, the gear 16 is provided by the central part of the second substrate 5 in contact with the periphery of the axis of rotation 14, and the rack gear 17 which has unevenness of the gear 16 and the gearing unevenness on the flexible region main part 41 is formed in it.

If the rotating handle 2 is rotated, the gear 16 will rotate with rotation of the axis of rotation 14, and rotational movement of the gear 16 will be converted to the translation motion of the rack gear 17. Here, the two flexible region main parts 41 slide to a counter direction mutually by the rack gear 17, and the flexible region 9 continuously formed in the flexible region main part 41 performs translation motion slid toward the central direction of the second substrate 5. As a result, the empty can compression nail 11 of the empty can pressing part 8a presses and compresses the empty can 19, after it moves toward the central direction of the empty can 19 and the point 12 of the empty can compression nail 11 contacts the periphery of the empty can 19.

[0031]

The four notches 43 are provided by the outer peripheral part of the second substrate 5. When this notch 43 unites the two empty can compressor implements 1, it is for storing the extending part 10 of the empty can pressing parts 8a and 8b of another empty can compressor implement 1.

[0032]

Fig.8 is a plan view of the third substrate 6.

As shown in Fig.8, the gear 16 is provided by the central part of the third substrate 6 in contact with the periphery of the axis of rotation 14, and the rack gear 17 which has unevenness of the gear 16 and the gearing unevenness on the flexible region main part 41 is formed in it.

If the rotating handle 2 is rotated, the gear 16 will rotate with rotation of the axis of rotation 14, and rotational movement of the gear 16 will be converted to the translation motion of the rack gear 17. Here, the two flexible region main parts 41 slide to a counter direction mutually by the rack gear 17, and the flexible region 9 continuously formed in the flexible region main part 41 performs translation motion slid toward the central direction of the third substrate 6. As a result, the empty can compression nail 11 of the empty can pressing part 8b presses and compresses the empty can 19, after it moves toward the central direction of the empty can 19 and the point 12 of the empty can compression nail 11 contacts the periphery of the empty can 19.

[0033]

Thus, have the substantially same structure as the second substrate 5, it is made to rotate 90 degree to the second substrate 5, and the third substrate 6 is laminated. Therefore, if the rotating handle 2 is rotated, the flexible region 9 attached to the third substrate 6 will slide in the direction perpendicular to the flexible region 9 attached to the second substrate 5. As a result, the empty can pressing parts 8a and 8b shown in Fig.1 and Fig.2 come to press the empty can 19 from the four quarters.

It is a direction perpendicular to the flexible region main part 41, and the two notches 42 are provided toward the central part of the third substrate 6 by the outer peripheral part of the third substrate 6. This notch 42 is interlocked with the flexible region 9 provided by the second substrate 5 shown in Fig.7, and functions as a slide guide at the time of the empty can pressing part 8a sliding.

[0034]

The four notches 43 are provided by the outer peripheral part of the third substrate 6 between the flexible region 9 and the notch 42. When this notch 43 unites the two empty can compressor implements 1, it is for storing the extending part 10 of the empty can pressing parts 8a and 8b of another empty can compressor implement 1.

[0035]

Fig.9 and Fig.10 are the plan views of the fourth substrate 7.Among these, Fig.9 is used for the empty can compressor implement 1 arranged above an empty can, and Fig.10 is used for the empty can compressor implement 1 arranged below an empty can.

As shown in Fig.9 and Fig.10, the hole 51 for letting the axis of rotation 14 pass is provided by the central part of the fourth substrate 7.On the fourth substrate 7, the fixed wall 13 for fixing an empty can is formed. When the empty can 19 is arranged and twisted, the holding part 52 for fixing stably, without the empty can 19 being slippery is provided by inner circumference slippage of the fixed wall 13. The holding part 52 can be formed in the surface by providing minute unevenness. This portion may be formed for the large material of sliding friction. Or when the empty can 19 is arranged and twisted, as long as it is a means for fixing stably, without the empty can 19 being slippery, it may be a thing of other structures.

When what shows an empty can to Fig.9 since the path is extracted with the up side forms the fixed wall 13 in main slippage rather than what is shown in Fig.10, the fixed wall 13 is tending to fix the side surface above an empty can.

[0036]

The four notches 53 are provided toward the central part from the outer peripheral part of the fourth substrate 7. This notch 53 functions as a slide guide at the time of the empty can pressing parts 8a and 8b sliding. The four notches 54 are provided toward the central part among the four notches 53 from the outer peripheral part of the fourth substrate 7. When this notch 54 unites at the time of non-use and keeps the two empty can compressor implements 1, it is for storing the empty can pressing parts 8a and 8b of another empty can compressor implement 1.

[0037]

Based on Fig.14, the directions for the empty can compressor implement 1 are described from Fig.11.

The rotating handle 2 of one empty can compressor implement 1 is turned down, and is arranged, the bottom surface of the empty can 19 is fixed on this empty can compressor implement 1, and, as for Fig.11, signs that turned another empty can compressor implement 1 up, and the rotating handle 2 is arranged for it to the upper end side of the empty can 19 are shown.

[0038]

Fig.12 shows the situation where the rotating handle 2 of the two empty can compressor implements 1 is rotated in the mutually different direction, and the empty can compression nail 11 of the empty can pressing parts 8a and 8b presses the side surface of the empty can 19 with rotation of the rotating handle 2. As a result, the pressed part is compressed and a dent produces the empty can 19 on the side surface of the empty can 19.

[0039]

Fig.13 is the figure which looked at signs that the side surface of the empty can 19 was compressed by the empty can compressor implement 1, from the upper part of the empty can 19, the empty can compression nail 11 pressed the side surface of the empty can 19, and the dent has produced it in the empty can 19 which was cylindrical shape.

The empty can compression nail 11 of the two empty can pressing parts 8a, The empty can 19 is contacted in the position which deviated from on each extension wire to the transverse direction for a while, and similarly, it is the position which deviated from on each extension wire to the transverse direction for a while, and in order to contact the empty can 19, as for the empty can compression nail 11 of the two empty can pressing parts 8b, the circumferentially regular dent of the empty can 19 is formed. It comes to end with the process of torsion of next by small power by making form of the empty can 19 into the form which has such a dent from cylindrical shape.

[0040]

Fig.14 shows the state where had the rotating handle 2 further and the empty can 19 was twisted, after stopping the slide of the empty can pressing parts 8a and 8b to the central direction of the empty can 19. The niting wall 24 mentioned above is realized by abutting the niting 27, and the stop of the slide of the empty can pressing parts 8a and 8b can perform two different operations by making this into a turning point by a series of continuous operation called rotation of the rotating handle 2. In order that the holding part 52 may contact Kamitsura of the empty can 19, and a bottom surface, in the case of operation of this torsion, can prevent the empty can 19 from racing, and at it the upper part of the empty can 19, Rotating united with the empty can compressor implement 1 installed in the upper part, the lower part of the empty can 19 rotates united with the empty can compressor implement 1 installed in the bottom. As a result, the empty can 19 is compressed by the vertical direction for a short time.

Since the empty can 19 is twisted after the circumferentially regular dent of the empty can 19 is formed, the empty can 19 which was twisted and was compressed by the vertical direction is prepared by the beautiful form, and it is convenient at Kami who puts and carries.

[0041]

In order to twist and compress the empty can 19, with cylindrical shape maintained, need large force, but. Since operation to twist will be performed after pressing the side surface of the empty can 19 and producing a dent if the empty can compressor implement 1 is used, the empty can 19 can be efficiently compressed into a vertical direction, without needing large force. Although it is the same as the thing of the description to a Patent document 1 about the point of making the side surface of the empty can 19 producing a dent, In the thing of the description to a Patent document 1, the ball attached to the rope crushes a can wall to the oblique direction where a rope twines, it is required for the rope to have slackened and to have stretched for that purpose, that there is nothing, and a problem is produced in respect of durability by this. On the other hand, the empty can compressor implement 1 of the present invention is making the side surface of the empty can 19 produce a dent with the slide of the empty can pressing parts 8a and 8b, and stable operation is guaranteed.

[0042]

Since the ball attached to the rope by inserting an empty can between an upper base material and a lower base material, and rotating both base materials to an opposite side in the thing of the description to a Patent document 1 is what crushes a can wall, Power required to compress an empty can is obtained after all with the power which twists an upper base material and a lower base material, and large force is comparatively required for it. On the other hand, in the empty can compressor implement 1 of the present invention, only by rotation of the rotating handle 2, operation which produces a dent, and operation which twists the empty can 19 from the upper and lower sides can be realized on the side surface of the empty can 19, and large force is not needed for it.

[0043]

Since the path of the rotating handle 2 is larger than the path of the empty can 19, even if it makes small power of rotating the rotating handle 2, it can obtain a big rotational torque and is advantageous also about this point. Rotation of the rotating handle 2 can perform all operations for compressing the empty can 19, and while operation is easy, time required for operation also ends extremely for a short time.

In order to remove the compressed empty can 19, what is necessary is just to rotate the rotating handle 2 to an opposite direction, and the operation in this case is also easy. In the thing of the description to a Patent document 1, after an empty can is compressed, it is in the state where it twined round the empty can by which the rope with a ball was crushed, and is hard to take out an empty can.

[0044]

Fig.15 shows the storage state at the time of non-use.

The rotating handle 2 of one empty can compressor implement 1a is turned up, and the rotating handle 2 of the empty can compressor implement 1b of another side is turned down, and as shown in Fig.15, if it unites and the two empty can compressor implements 1 are kept, a place is not taken, but it can save, without losing one side. Although it has the structure projected toward the center of a radial direction, the empty can compression nail 11 of the empty can pressing parts 8a and 8b, Since the notch 32 provided by the first substrate 4 is formed so that it may have only the depth which can store the empty can compression nail 11 when the two empty can compressor implements 1 are united as it described

based on Fig.6, The rotating handle 2 can be rotated suitably and the empty can compression nail 11 can be stored to the notch 32. Therefore, when the two empty can compressor implements 1 are united, it can be considered as the compact form which makes cylindrical shape substantially as a whole.
[0045]
As it described based on Fig.6, rather than the notch 43 shown in Fig.7 and Fig.8, the depth is deep toward the central direction and the storage of the empty can compression nail 11 projected to a central direction by this of the notch 32 provided by the first substrate 4 is attained.If the empty can compression nail 11 is stored by the notch 32, since it inserts into the rotating handle supporting part 3 and the second substrate 5 and is crowded, the empty can compression nail 11 will be fixed just.
[0046]
In Fig.15, the empty can compression nail 11 of the empty can pressing part eight a1 of the empty can compressor implement 1a arranged at the upper part, It is stored by the notch 32 provided by the first substrate 4 of the empty can compressor implement 1b arranged at the bottom, and the empty can compression nail 11 of the empty can pressing part eight a2 of the empty can compressor implement 1b arranged at the bottom is stored by the notch 32 provided by the first substrate 4 of the empty can compressor implement 1a arranged at the upper part. The same storage is made about other empty can compression nails 11. Thereby, even if the two empty can compressor implements 1a and 1b are required in compressing the empty can 19, at the time of non-use, it can unite and these can really be made into a thing. Since it can be kept in such the state, [in a home, a restaurant, or a car etc.], it can be kept without taking a place, and when an empty can is made, it can be used easily. If the rotating handle 2 is rotated to use the empty can compressor implements 1a and 1b in such a storage state, the empty can compression nail 11 accommodated in the notch 32 will slide to an outer peripheral direction, and a union state will be released.
[0047]
In the thing of the description to a Patent document 1, although an upper base material and a lower base material are connected with a rope and it has really

become a thing, at the time of non-use, inevitably, it exists, after the rope with which the ball accompanied has slackened between an upper base material and a lower base material, and cannot be kept as a fixed form thing of the fixed form. Therefore, when taking out from a storage place, a rope twines, or a rope is caught in other things, and it is use top inconvenience. On the other hand, in spite of using the two empty can compressor implements 1 by a pair in the empty can compressor implement 1 of the present invention, it can unite, the two empty can compressor implements 1 can be kept as a fixed form thing, and convenience is high also in the case of storage.
[Industrial applicability]
[0048]
The present invention is easy to operate and can be used as an empty can compressor implement which is rich in durability while an empty can is easily compressible by small power. In particular, in a home, a restaurant, etc., after drinking beer and juice, in order to discard a used empty can, it can use as an empty can compressor implement which can be used easily.
[Explanations of letters or numerals]
[0049]
1, 1a, 1b empty can compressor implement
2 Rotating handle
3 Rotating handle supporting part
4 The first substrate
5 The second substrate
6 The third substrate
7 The fourth substrate
8a, 8b, eight a1, an 8a2 empty-can pressing part
9 Flexible region
10 Extending part
11 Empty can compression nail
12 Point
13 Fixed wall
14 Axis of rotation
15 Washer

16 Gear

17 Rack gear

18 Pull tab

19 Empty can

20 Body part

21 Engaging part

22 Axis-of-rotation receptacle

23, 23a, and 23b Wall

24 Niting wall

25 Hole

26 Engaging part

27 Niting

31 Hole

32 Notch

41 Flexible region main part

42 Notch

43 Notch

51 Hole

52 Holding part

53 Notch

54 Notch

空き缶圧縮ビジネス 空き缶圧縮器具の使用方法

定価（本体１,０００円＋税）

２０１２年（平成２４年）１１月５日発行

No. TA-024

発行所　発明開発連合会®
東京都渋谷区渋谷 2-2-13
電話 03-3498-0751㈹
発行人　ましば寿一

Printed in Japan
著者　田中勝之 ©

本書の一部または全部を無断で複写、複製、転載、データーファイル化することを禁じています。
It forbids a copy, a duplicate, reproduction, and forming a data file for some or all of this book without notice.